Tarka Country *follows the story of Henry Williamson's famous otter "Tarka" and "his joyful water-life and death in the Country of the Two Rivers."*

North Devon naturalist Trevor Beer, takes us on a journey, visiting all Tarka's favourite haunts and scenes mentioned in the novel "Tarka".

His book also includes notes on the wild animals and birds featured in the story.

Cover photo of Canal Bridge, River Torridge by Trevor Beer

Photo on Page 3, River Torridge by Sandra Yeo

Tarka Country

TREVOR BEER

The publishers wish to acknowledge the permission granted by The Bodley Head, Covent Garden, London, to quote from Henry Williamson's "Tarka".

Photography and illustrations by the author except where otherwise stated.

First published in 1983 by Badger Books, Bideford, North Devon.

Typeset by Lens Typesetting, Bideford, North Devon.

Printed in Great Britain by Devon Print Group of Exeter.

INTRODUCTION

Tarka country – the fabulous countryside of the two rivers, Taw and Torridge – the beautiful countryside followed by otters and the many other wild creatures described by Henry Williamson in his famous nature story "Tarka The Otter".

Times have changed since those days. A half century or so and mankind has made rapid progress, or strides, yet the land itself has altered little.

True, towns and villages have grown, many woods and hedgerows that Williamson knew are gone, but North Devon is still relatively unspoilt, still not too intensively farmed.

Certainly it is the wildness, the relatively unpolluted waterways with good cover and some seclusion, that has held our otter population at a healthy level and, too, the county's flora and fauna generally.

This book followed the tracks laid down by Tarka and his kind since those early days. Rainwashed away they may have been many thousand of times yet still they reappear.

I decided the only way to write "Tarka Country" was to follow the trails laid down by Henry Williamson, the trails he definitely followed himself for they are accurate, and the wildlife found along them is also accurately described.

Each area has been visited – often by night and day and in the seasons when Tarka was there. This was a joy in itself for these are the haunts of the naturalist and country lover.

Landowners were and are helpful – my thanks to them all – but I must remind readers that many places have no public access. There are, however, numerous vantage points from which we can view Tarka country perfectly well.

One thing's for sure, Henry Williamson was there. Having walked, climbed, slid, scrabbled and splashed in every one of these places myself, I know he wrote from experience.

If that reads like a tribute to the man, then so be it . . .

River Taw *Sandra Yeo*

River Torridge

Trevor Beer is a naturalist and wildlife artist and lecturer born in North Devon. In 1979 he won the Gavin Maxwell Award for his work on the conservation of otters, an Award given every few years by a panel of wildlife experts, Gavin Maxwell being the naturalist and author of the otter story "Ring of Bright Water".

He is grateful to all those who have helped him on his journeys over the Two River Country and the kindness shown him by many landowners.

HENRY WILLIAMSON was born in 1896 and at the age of seventeen found himself soldiering in the trenches at Ypres in the autumn of 1914. On Christmas Day of that year he was present at the fraternisation between the English and German troops, this extraordinary event changing his outlook on life. After the War, Henry left London and came to North Devon where he rented a tiny cottage, cob-built and with Barn Owls nesting in the thatch. The Barn Owl became his symbol, painted on his front door and used as the sign manual to his many books. Henry lived a lone life in the cottage for several years, tramping the countryside and living with his dogs, cats, wild birds and, at one stage, an otter cub which he rescued from a holt after the mother had been shot.

The otter cub refused Williamson's fountain pen refill substitute teat and he eventually persuaded his cat, already nursing a kitten, to suckle the young otter also. The cub thrived, feeding at first on milk then on fish, rabbit, dog biscuits and vegetables, wandering the countryside with Williamson and coming readily to his call. On one such evening walk the otter was caught by a paw in a rabbit gin, almost severing three toes in its frenzy to escape the terrible pain.

Williamson managed to release the struggling otter but the frightened animal wriggled from his grasp and disappeared. Williamson searched and called for weeks, wandering far afield along the coast, the estuary of the Two Rivers – the Taw and Torridge – following them to their very sources in an attempt to find his beloved otter, but in vain.

But in Williamson's mind, the story of "Tarka The Otter" was born. More at home than ever with the wildlife of North Devon and now knowing his countryside extremely well, he even went out with the Otter Hunt to find out for himself their side of things, ever seeking the truth from actual experience.

During the time he was writing "Tarka" he married and a son was born. The family remained in the cottage, Mrs. Williamson fighting poor health whilst her husband helped care for the family, cooking, cleaning and continuing with his book. The book was finally finished and published in 1927.

Henry Williamson continued to live in North Devon, eventually buying a much larger country residence at Ox's Cross, near Georgeham, which has since been sold out of the family. But the famous writing hut remains in the grounds and is a place of pilgrimage for Henry Williamson followers. Built of elmwood it has been restored by the Henry Williamson Society, an organisation devoted to promoting the works of the author as one of the great writers of the twentieth century.

Henry was, of course, more than a writer. He farmed and was a practical conservationist, building barns, renovating cottages, making roads even, a man who wrote with a full knowledge of what he was writing about.

He died in 1976 during the filming of "Tarka The Otter" in North Devon. His death occurred as the actual scene of Tarka's fight to the end with the hound Deadlock was being shot, and he never saw the finished film that has delighted so many.

Twilight over meadow and water, the eve-star shining above the hill, and Old Nog the heron crying 'kra-a-ark'! as his slow dark wings carried him down to the estuary. A whiteness drifting above the sere reeds of the riverside, for the owl had flown from under the middle arch of the stone bridge that once had carried the canal across the river.

THE RIVER TORRIDGE

Tarka the otter was born on the Torridge.

Henry Williamson chose this, the home of otters for thousands of years, as the true beginning of his story.

What better choice? ... And one which has stood the test of time, for despite the drastic decline in their numbers in Britain as a whole, otters survive in North Devon. Tarka "the little water wanderer" took his name partly from the Celtic "Ta". Williamson explains that this word means water and is the probable derivation of the river names the Taw and the Torridge.

Thus was the delightful name "Tarka" born, and the otter himself in a holt not far from Canal Bridge, an attractive spot between Bideford and Torrington, no way marred by the man-made structure. In fact the bridges of the Two River country enhance the waterways they span.

The twelve great trees Henry Williamson refers to, he tells us there were once thirteen, have all disappeared since Tarka was written. Now, in the 1980s, thirteen young trees have been planted in their place, marking the dedicated work of the Otter Haven Project in North Devon, as well as the ready co-operation of farming landowners.

Quite different from the Taw, the Torridge is far more wooded for much of its length. This extra cover and seclusion provides ideal otter habitat which greatly aids their survival.

Not a place to be opened up for tourists, but a place to be maintained and protected in order that visitors and locals alike can appreciate its naturally green and tree-lined banks. A place, then, to be respected as a natural wildlife sanctuary where the public is encouraged to pause, view and enjoy and to move on, leaving the waterway safe for otters and other wildlife.

The Torridge differs from the Taw also in that it does not rise on Dartmoor in its own right. One of its principal tributaries, the Okement river, rises near the Taw on wildest Dartmoor, giving name to Okehampton town, an excellent place from which to visit this part of Tarka Country.

Landcross

Weare Giffard *Sandra Yeo*

TARKA'S BIRTHPLACE

Tarka was born, the eldest of three cubs, at Owlery Holt near Canal Bridge and here he and his two sisters would have spent the first six or eight weeks of their lives before venturing out into the great wide world. We read of their weaning, of the first fish they eat and of kingfishers, voles and the like. Tarka's first solid food was eel, the otters' favoured diet when available.

Tarka learnt to swim near Willow Island, Peal Rock and the Canal Bridge, enchanting places with equally enchanting names, and, too, the Moon Field and Burnt Sycamore Holt.

Here in his cubhood days Tarka met with nightjars, "whose two eggs were laid among ferns in the wood". A rare sight today, nightjars nesting in North Devon. It took me three years to find the species on the Torridge in a breeding situation. It is at Owlery Holt that Tarka first learns of the Hunt and of Deadlock the Otterhound, the hound that would plague him throughout his life.

And on this day Tarka's father was killed by the Hunt as the young cub lay safe with his mother and sisters in the holt.

Shortly after this incident mother and cubs moved away via Upalong and Waymoor, wildly beautiful countryside leading to the Claypits of Marland.

Marland Moor is a lovely area. The old ponds are still there and the countryside is surprisingly wild.

Henry Williamson's writings came to life here. Dragonflies zoomed about in abundance, Golden Ringed Hawker, Southern Aeshnae and the Large Red Damselfly, all were there adding to the beauty and colour of the place which has an atmosphere of remoteness and peace. Marland was once Merland, by the Mereside, and is well worth a visit, a must for Tarka fans.

Nearby Huish, once Hewisch, gave its name to several lords who held the land, a Philip Huish holding land here in the reign of Henry the Second. History also tells us that Emma Huish, last of this line, was the wife of Sir Robert Tresihan, a Chief Justice of England.

As to Huntshaw, Shaw in the olden days was Shadow and thus Huntshaw was Hunters Shadow, an appropriate name for an area well wooded. The Huntshaw Water sees the occasional otter to this day — but more often the all too frequent mink.

Mere Brook and the Little Mere river run away across country from the road below Rosehill and Huish. Here hounds hunted Marland Jimmy, the old dog otter, and we learn of an old hunt "trick", the pouring of paraffin into the waters to bring an otter out of hiding.

Williamson tells of the "gins tilled" for the enemies of the young pheasants in the woods, the "enemies" being the natural predators, stoat, weasel and hawk, not man who rears the pheasants to shoot them at a later stage. Even today gamekeepers' gibbets are hung with the dead creatures considered harmful to game rearing. Hopefully today's gamekeepers are a more enlightened breed than those in Williamson's days, for he refers to dwarf owls (the little owl), buzzard, kestrel and sparrowhawk, all of whom rarely, if ever, take game. The little owl, for example, is virtually an insect eater.

The Hunt on this occasion was unsuccessful and Marland Jimmy spent the rest of the day ridding himself of the stench of paraffin, Tarka moving on with his mother and sisters to Braund's Hill Wood and the Torridge. Henry Williamson's references to the old otter, Marland Jimmy, are of an otter very much a nomad and becoming sedentary in his old age. This is very much the case with otters generally and old otters' territories tend to be small, chosen where ample food can be obtained with relative ease.

Marland Jimmy was a traveller beyond compare in his early days. Salmon killed in the Severn, pollock eaten on the rocks at Portland Bill and lampreys in the Exe, hundreds of miles travelled in his lifetime before coming to the white claypits of Marland to end his days.

Just how accurate such long distance travels are we cannot say. One otter with its coat splashed with yellow paint was seen by local anglers in many parts of North Devon a few years ago. Easily recognisable, he came off the Torridge to Instow and thence to Fremington and via Barnstaple to the "shelleys" beyond the town near Bishops Tawton. He was then seen near Croyde some months later, so got around a bit.

BRAUND'S HILL WOOD

It was here that Tarka had his first encounter with the terrible traps men fashion to maim and kill wild creatures.

The otters, attracted by the sounds of fighting stoats, smelled the flesh of rabbit hidden in a pipe, the cubs' curiosity getting the better of their caution. Tarka's sister, the youngest, beat him to the tantalising scent and was caught by the metal gin.

It was in the ensuing battle with gamekeeper, retriever and the gin trap itself that Williamson portrays the courage and devotion of an otter bitch for her cubs. She is no match, however, for the shotgun fired by the keeper and with her youngest cub shot dead in the trap, she and the remaining cubs are forced to flee.

Tarka, himself bruised by the snapping shut of the trap, is consoled by his mother, sleeping with his sister after a meal of moorhen's eggs in a hover by the crossing pool.

For many days afterwards, the otters enjoy the Torridge river country thereabouts, fishing for brown trout to the sounds of Crackeys, Ackymals and Ruddocks, the Devon names for wren, blue tit and robin as Williamson tells us.

The otters spend happy days from the Crossing Pool to Golden Pool, journeying towards Autumn via Canal Bridge and Rothern Bridge to lovely Beam.

They killed rabbits and met Stikkersee the Weasel who raves at their being in his woods.

At Beam Tarka meets with the salmon poachers hunting with a pitchfork to prang fish seen by oilrag torchlight in the pools. In the darkness one poacher runs a pitchfork prong into his hand and still struggling to land the fish is bitten in the leg by Tarka. The poacher's dog is also bitten by "the little brown dog" and a night's poaching is somewhat thwarted.

Since Williamson's days, poaching has increased. As well as the old style poacher taking game for his family table, there are the mercenary gangs using ruthless, indiscriminate methods who travel far afield for salmon, even hundreds of miles, when the price is right. The River Wardens of the South West Authority have their work cut out.

Beam Weir

Rothern Bridge

While swimming in this happy way, he noticed the moon. It danced on the water just before his nose. Often he had seen the moon, just outside the hollow tree, and had tried to touch it with a paw. Now he tried to bite it, but it swam away from him. He chased it. It wriggled like a silver fish and he followed to the sedges on the far bank of the river, but it no longer wriggled. It was waiting to play with him.

Close by Beam are Weare Giffard and Annery, lovely countryside where the Torridge may flow peacefully, or suddenly become a raging torrent after heavy rains. Look for flood debris hanging halfway up the trees along the riverbanks here to realise just how ravaging these stormy times may be and the effect they must have on otter haunts. Here is some of the most beautiful of the Torridge scenery as well as much of interest for the naturalist. Scores of bird species from Britain's smallest, the goldcrest, to the larger mute swans and grey herons are here, along with a rich flora and many mammals and insects.

Weare Giffard lies low beside the Torridge, taking its name from the nearby weir and was the ancient inheritance of the Giffard family, one of whom, Sir Walter Giffard, lived here in the reign of Henry the Third.

Notable North Devon families mixed and matched in those days, the Giffards and Fortescues linked by marriage to each other, the Rolles of Heanton and the Chichesters. It was while exploring otter country in this area I met with Mrs. Lampard-Vachell, widow of the late Edwin Lampard-Vachell, a great local benefactor. She still resides at Weare Giffard and we discussed the olden days of otter hunting and also her late husband's well known book, "The Wild Birds of Torrington & District".

When the moon had come to its full round shine, Tarka was hunting his own food in the pools and necks of the clear water running round the bend above Canal Bridge, which rod-and-line men declare to be the best salmon beat of the Two Rivers.

Few public footpaths are shown on OS Maps of the area, and much of the land is private. There are footpaths around the golf course just south of Furzebeam Hill, and

Weare Giffard Hall   *Sandra Yeo*

Salmon netsmen at Bideford Bridge.

again at Weare Giffard in the area of the church, and eastwards to the road north to Gammaton Moor.

However, there is no problem viewing the river from the roadsides and some of the finer views of the Torridge are to be found between Beam and Pillmouth where one of the many Devon Yeo's joins the main river system.

Landcross Bridge is an ideal spot from which to view the river and its wildlife. Here it is the river Yeo flowing beneath the bridge, with the Torridge to the north.

Shelduck, mallard and a variety of wading birds can be seen on the mudflats from the bridge, a good place to pause awhile and enjoy Tarka Country.

From Beam Weir Tarka, his mother and sister move on via Never-be-good Wood to Brimacombe Brake. It is here that he first meets White-Tip and Greymuzzle, the two bitch otters both of whom are destined to become his mates. They travel on towards Bideford, at one time sleeping in the sett of badgers, then on again via "Lancarse Pill", (Landcross) where White-Tip was born, until they eventually reach Bideford Longbridge.

The five otters swim beneath the Longbridge as the traffic rumbles overhead.

Williamson tells of the bridge built by monks "two centuries before the galleons were laid down in the shipyards below to fight the Spanish Armada". Recently re-structured because it had a touch of the wobbles, the bridge has twenty-four pointed arches and was supposedly built following a vision which came to Sir Richard Gornard, a priest of the town. The ancient stone structure with its pointed arches has been left intact.

Bideford, By-the-ford, is well known for its seafaring locals of past and present. Probably the best known family is Grenville, a Sir Richard Grenville, a Knight in the reign of William Rufus renowned for his valour. Another Sir Richard Grenville in the reign of Queen Elizabeth the First fought the greatest ever sea battle by an Englishman against the Spanish Fleet.

The Domesday Survey records the value of Bideford as a fishing port, the town retaining much of its character despite a veneer of modern so-called improvements.

But such are not the concern of otters and Tarka and his four companions left the Torridge for a while to visit the Braunton Pill and Ramshorn Pond as swallows were congregating amongst the reedmace before following the sun to Africa . . .

Tarka was alone, a young male of a ferocious and persecuted tribe whose only friends, except the Spirit that made it, were its enemies – the otter-hunters. His cubhood was ended, and now indeed did his name fit his life, for he was a wanderer, and homeless, with nearly every man and dog against him.

Ramshorn Duckpond is part of the vast area of Braunton's marshes and Great Field that Williamson refers to as the "great plain". "Branton" is Braunton, of course, and Williamson makes good use of many Devon dialect pronunciations. Though not a Devon born man he picked up the dialect and used it freely when in the mood, as a native.

Here at Braunton ponds, Tarka is at last driven away from his mother's side by her newfound mate. Tarka is independent. Strangely we do not really know what becomes of Tarka's remaining sister. Their mother chased them both away when her mate-to-be paid the younger female too much attention. We must assume both cubs became loners at this time. Tarka moved to Braunton Pill and came to the power station and the railway.

Readers unfamiliar with the area should not confuse the reference to the power station with the huge East Yelland Power Station complex just across the estuary near Instow. The power station Tarka knew has long since disappeared and only a few signs remain, though the man who last worked the station still resides in Braunton village.

Already the Yelland station, built in the 1950s, is obsolete and may shortly close. The railway, too, is disused. This, the one time route to Ilfracombe from Barnstaple, is now a public footpath for much of its length and can be walked from Barnstaple to Braunton giving fine views of Tarka country.

On the River Caen beyond Braunton, Tarka again finds White-Tip, the bitch otter. They play in Buckland Meadow and we learn that White-Tip's mother had been killed by the Otter Hunt in late September, and that White-Tip had been parted from her friend Greymuzzle when a Marshman's dog chased them. Now, as they play along the riverbank, a dog otter from Exmoor arrives and Tarka loses the fight for White-Tip's favours to retire battle-scarred and alone yet again.

Back to the small power station he goes, sliding over the weir "like oil", riding the tide out to the estuary where

River Caen, Braunton Marsh *Sandra Yeo*

he hears the whistle of an otter. It is Greymuzzle who licks his wounds and they hunt together, "and in the course of time she takes him for her mate" . . . They sleep in the day-hide of a bittern, spending some weeks here at Braunton before moving on to Baggy Point near the village of Croyde.

BAGGY . . .

The magnificent headland of Baggy is reached by Tarka and Greymuzzle after a five mile swim along the shallow coast. A well-maintained coastal footpath enables us to reach Baggy more easily and on any day, be it rainy or sunny, the views are superb.

The screaming of seabirds and jackdaws above the sound of the waves crashing three hundred feet below upon the jagged rocks is far removed from the green, shady river reaches, a breath-taking place.

Here dwelt Jarrk the Seal, hunting conger at Bag Leap and playing in the huge cave known as Seal Cavern where these beautiful creatures may still be seen in play today. There are several caves about the coast, dark and eerie places full of strange shapes and shadows, moisture dripping from the roofs with that amplified metallic sound of all such places. Williamson tells us that many seals died in the Seal Cavern, referring to bones, skulls and shrunken hides. In his day the seals were more common, though they may still be seen basking on Lundy's beaches and breed on the North Cornish coast. Conger are commonly caught from Baggy and about the headland, fishermen favouring this bit of coast around to Sandy Cove and Long Rock.

The Cormorants Rock, aptly named for cormorants and shags regularly perch upon it, can be viewed from the coast path. Climbing the Wreckers Path, and exploring the caves and coves is to be in a place of peace with only the seabirds and a few rabbits.

The area is the haunt of geologists who come for the fascinating visual evidence of a raised beach. The raised beach section stretches from the northern end of Baggy Point for some three miles southwards and is said to be one of the finest known examples of its kind.

About half a mile west of Saunton Hotel is a large erratic red granite boulder resting on the raised beach platform and embedded in the sands above high water mark. Said to have derived from Scotland the boulder measures about 7 by 5 by 2 feet. Another erratic of a greywhite colour and about 18 cu. ft. in size lies on the foreshore of the south side of Croyde Bay.

Nearer Saunton I found the ancient nest-site of Kronk the Raven, an exciting find for ravens still nest in the very spot! To know that Tarka Country exists, in the main, just as it was when Henry Williamson wrote his story is in a way very reassuring – the large stick nest with its sheeps' wool lining, the shining black plumage of the young ravens, the deep throated call of the adults, all so beautiful as the setting sun dipped redly behind them.

Eggs laid in the stormy month of March on the precarious sandstone cliff, the young growing fast to fledge in the gorsegold and pink thrift days of spring, such is the wild life of the raven as it reigns supreme amongst jackdaws, stonechats, rock pipits and other bird species nesting nearby. And Tarka and Greymuzzle – their happy times are eventually disturbed by human presence at Baggy and once again they are on the move back towards Braunton and Horsey Marsh.

BRAUNTON'S "GREAT PLAIN"

Henry Williamson was in all probability referring partly to the Braunton Great Field, though references are brief and the story is more of the terrible winter conditions and the arrival of many otters – families of three and four, one of five otters. White-tip and her mate, Greymuzzle, with a cub, and Marland Jimmy, some twenty or more otters, all driven there by the extreme weather conditions.

The Great Field, just south west of the village, is one of the few remaining examples of the ancient strip system of farmland tenure. This communal system of farming is medieval in origin and was documented in Braunton's records over 600 years ago.

The Great Field is about 350 acres.

South of the Great Field lies Braunton Marsh, an extensive area of reclaimed saltmarshes cut off from the sea by the huge bank built in the nineteenth century. Numerous freshwater drainage dykes dissect the marshland and the whole area is rich in wildlife.

Horsey Island, still further south, is also reclaimed land. "Island" because it is separated from the mainland by a tidal creek following the former channel of the Caen river. It is possible to view over this whole "Great Plain" area from Braunton's West Hill, said to be one of the most beautiful and interesting views in the whole country. The land reclamation was carried out between 1811 and 1815 by local labour augmented by Dutch, Irish and Cornish workmen and all at a cost of £20,000.

Though some arable crops were grown, the land eventually became cattle grazing country recognised as some of the best in England. The Horsey Island project

The Estuary from Braunton   *Sandra Yeo*

Sandra Yeo

came later, in the 1850s, the final gap being closed to the tides in 1857 "by 320 men with 140 carts". The facing stones of the sea wall came mainly from Braunton Down Quarry with some boulders being taken from the estuary.

Ruined linhays and barns are scattered about the marshes, allowed to fall into disrepair when lands reverted to the landlords from tenant farmers. This unfortunate state of affairs has somewhat despoiled the landscape, as well as depriving cattle of shelter and barn owls and other wild creatures of some useful accommodation. As recently as the 1950s and 60s the area had a much better kept appearance.

Hereabouts I would meet up with Henry Williamson, chatting while the "rusty reed-daggers shook their fist heads at the sky" as he himself would do when Chivenor jets roared overhead. It was cold then and Williamson's eyes would water hugely in the biting winds, his heavy dark coat pulled about the shoulders normally straight, but hunched now to the winds as teal and snipe put up about us. Such meetings were accidental, time after time the marshes would be empty of people on almost every visit, and then the lone figure would appear, or in more recent years before his death, be driven along the marsh roads in a huge black car. And on these marshes the "Tarka" story lives on ... each winter the "sedge owls" arrive from the north, these the short-eared owls which visit us annually and being a diurnal species are seen about the marshes by day.

Williamson refers also to "Crow Island" at the edge of Braunton Burrows as "a spit of gravel crowned by sandhills and bound by marram grass". He would have remembered those days well and, though Crow Point as we know it today is part of the mainland, it was certainly an island in the 1920s. There are people alive today who recall that boats sailed between Crow Island and the beach and today it seems possible the land will revert once more to an island. The seas are again threatening to breach the area and it seems probable that Crow has been an "on and off" island for centuries.

Tarka and Greymuzzle found the Braunton area beckoned them with the icy fingers of what was to become a bleakly severe winter. All was white with frost and ice, a terrible winter throughout Europe, for Williamson tells of an Arctic Owl, White-fronted Geese and a Greenland Falcon travelling "before the blizzard howling its way from the North Star"...

When such birds are seen in North Devon they are driven here by far harsher conditions elsewhere, seeking refuge from the bitter, foodless wastes of arctic conditions. The rule for North Devon birdwatchers is the harder the winter the more waders and wildfowl we see.

To Greymuzzle and Tarka times were made harder by the birth of their single cub at a time when finding food even for themselves became desperate.

Greymuzzle nursed the cub on the marshes, "the heat of her body melting the snowflakes", the poor animal frost-bitten and blinded by the terrible conditions. The flatfish moved out over the bar to warmer waters and "even the Crows died of starvation" as did Old Nog the heron's mate, and Marland Jimmy, the old dog otter from the claypits of Marland Moor.

Tarka, hunting at Scur Farmyard, was caught by one leg in a gin trap.

Greymuzzle, hearing the noise of the springing of the trap came to his aid, "gnawing in fury the iron jaws of the gin". Fighting the trap she bit through the sinews of her mate's foot and Tarka is freed, running to the river in fear and pain. Greymuzzle, remembering her starving cub, remains for she has heard ducks within a shed in the farmyard. Desperately gnawing at the wood of the shed door she frightens the ducks who in turn awaken the farm dog and the farmer.

Williamson tells of the farmer examining his gin trap and finding three toes from Tarka's paw and then of finding the two otters together, huddled in a shed for refuge.

Again Greymuzzle faithfully defends her badly injured mate, fighting the collie dog as Tarka escapes through a hole in the wall. Weakened by starvation Greymuzzle was not able to fight for long and she is held fast by a dungfork and killed with an iron ferreting bar.

Tarka "remained all night in the farmyard, waiting for the mate that never came", moving away "in the mist and rain of the day, to hide among the reeds of the marsh

Sandra Yeo

pond". Williamson tells us he walked there himself, searching for signs of Tarka . . . "I walked to the pond, and again I sought among the reeds, in vain; and to the pill I went, over the guts in the salt grey turf, to the trickling mud where the linnets were fluttering at the seeds of the glasswort. There I spurred an otter, but the tracks were old with tides, and worm castings sat in many. Every fourth seal was marred with two toes set deeper in the mud".

And we know that Tarka has left the Braunton marshes to begin a great journey to Dartmoor and the source of the Taw at Cranmere and Taw Head.

> *Tarka was gone in the mist and rain of the day, to hide among the reeds of the marsh pond – the sere and icicled reeds, which now could sink to their ancestral ooze and sleep, perchance to dream of sun-stored summers raising the green stems, of wind-shaken anthers dropping gold pollen over June's young maces, of seeds shaped and clasped and taught by the brown autumn mother. The south wind was breaking from the great roots the talons of the Icicle Spirit, and freeing ten thousand flying seeds in each brown head.*

THE TAW

The Taw, or "Gentleman's River" as it is referred to by Henry Williamson, rises on Dartmoor in as wild a spot as anyone could imagine.

Here, as on Exmoor, mists falls without warning, enveloping the countryside and the unwary traveller quite alarmingly. At such times one almost expects to see the legendary black hounds that haunt the moors, a legend that may well have given rise to Conan Doyle's "Hound of the Baskervilles," the Sherlock Holmes' tale set in the wilds of Dartmoor.

But why "Gentleman's River"? Williamson tells us it is the name given to the Taw by the Otter Hunters because the inns were so placed that one could get refreshments at lunch time wherever one called a halt. Such is the case today and several inns along the waterways offer excellent fare.

But the Otter Hunts are no more. Otter hunting is illegal though working hounds may still be seen hunting mink in the Two River country. Ironically perhaps, due to the banning of the "sport" the otter hound is said to be in decline, though most Hunts used a majority of foxhounds and rarely the true otter hounds anyway.

But let's away to Dartmoor and the source of the Taw, a place so wild that the tiny wading bird, the dunlin, still breeds in the desolate bog country that will suck the rubber boots off the legs of the unwary.

Cranmere is a bleak place, to some unattractive, but the beauty of Devon lies in its variety of habitat. Cranmere is differently beautiful, its bogs and mires and the marshy pool that is the source of several Devon rivers, a place of history and mystery. The granite and clay basins here are constantly fed by the moorland rains and mists. From the very pool of Cranmere rises the West Okement while nearby rises the East Okement, the Taw, the Teign, the Tavy and the Dart.

Generally speaking the watershed is southwards, all but the two Okements and the Taw flowing to the English Channel.

A place of legend is Cranmere, and the pool is said to be bottomless.

One legend is of a dead farmer whose ghost was so troublesome it took seven persons to secure it and with their pious spells they changed him into a colt. A young lad was told to lead the colt to Cranmere Pool, loose it from its halter and leave without a backward glance. He followed all but the last instruction and received a kick from the animal which then plunged into the pool in the form of a fireball . . .

Military use of this area has been going on since at least the beginning of this century, when in the early 1900s the Royal Horse and Field Artillery leased some ninety acres of the moor.

This then is Tarka Country at its wildest, probably its wettest, and any otter arriving here would soon away, back along the waterways to kinder fish-filled places.

Snipe may be seen here in the misty rains of late autumn, putting up from under ones feet to zig-zag away. In springtime they may be seen displaying at the nest area, sometimes calling from song perches, an unusual view of this lovely wader.

Close to Cranmere Pool is one of the famous Dartmoor letter boxes from which mail can be specially postmarked.

The trickle of water that is the Taw flows swiftly northwards passing disused mines and on via Oke Tor into the wild and lonely Taw marshes. Here Small Brook enters the river system from the east where there are some enclosed hut groups. Just to the west tracks lead to Winter Tor and Higher Tor, the Irishman's Wall and Belstone Tor. This whole area is beautiful, the Taw picking up other small tributaries eastwards from the marshy countryside around Foxes Holt.

A useful centre from which to visit this area of Tarka Country is Sticklepath with its hotel and inns in the village, and at Belstone itself. Okehampton is nearby to the west and time spent exploring Tarka's Dartmoor from hereabouts would not be wasted.

The river hurried round the base of the cleave, on whose slopes stunted trees grew, amid rocks, and scree that in falling had smashed the trunks and torn out the roots of willows, thorns and hollies. It wandered away from the moor, a proper river, with bridges, brooks, islands, and mills.

The Taw reaches Birchy Lake near Belstone, swinging sharply eastwards through Belstone Cleave and beneath the road at Sticklepath, where it again swings north as a rushing moorland waterway. Tarka ate rabbit here after a dispute over the kill with Swagdagger the Stoat near the Seal Stone.

Now the river is a sparkling, rushing moorland stream where that delightful Devon bird the dipper may be seen hunting insect life and tiny fish prey beneath the surface of the water. The currents, strong and often swollen by heavy moorland rains, are tested by trout and the occasional otter that still comes here.

Otters hunting this area along the dashing Dartmoor streams may well leave their spraints by ancient clapper bridges, the huge flat granite slabs supported by even thicker slabs. To know Tarka's Dartmoor one must follow its ancient trackways and watercourses as he did, on foot. The magic of the place entirely eludes the carbound who merely skim the surface of its beauty.

Belstone Cleave is a rare place with a Tor that rises some 1568ft (522m) in height. The term Tor is used only where the granite actually protrudes from the ground. Here too are the famous Nine Maidens' standing stones, the stone row leading to these, 'the Dancers', being all of two miles long. Nine girls were turned to stone as punishment for dancing on a Sunday, says the legend. Another fallen stone nearby is said to be their piper who was similarly punished. And they still dance here each day at noon – or so the story goes.

A pixie-laden place where even today some locals put out a bowl of milk to invoke the pixies' goodwill rather than their wrath. Once only did Henry Williamson refer to Dartmoor in our all too brief conversations.

"You must wear your coat inside-out up there" he said and went on about something else. It was long after that I discovered such inside-out apparel apparently appeases the pixies.

In the fading light of a summer's day when wraiths of mist swirl about the stones and along the paths and waterways, it is not difficult to understand why people feel a tingle along their spine and even dogs and horses tremble.

One of the hounds haunting the moor here is said to be the ghost of Lady Howard who murdered two of her four husbands at Okehampton Castle. It is said she spends her time in the form of a hound carrying blades of grass between Tavistock and Okehampton.

Near Cawsand Beacon is the dreaded Raybarrow Pool, a place only for the likes of Tarka for it is the most dangerous of all Dartmoor bogs.

Tarka moved away from Belstone, chased along the wooded cleave by stoats called to the scene by Swagdagger, angry at the otter stealing his rabbit kill. The whole pack chased Tarka to the Taw, three falling in after him to crawl out "spitting and sneezing".

Tarka followed the wandering river away from the moor towards Colleton, a long journey of many a changing scene. Williamson refers but little to the Taw until it reaches Colleton Mills, but we know Tarka passed on northwards from Sticklepath.

Close to Rat Combe small streams run into the river from the west, the Taw suddenly widening where it flows beneath the roadbridge at East Rowden, and yet again beneath the A3072 where I found otter spraints only a day or so old in June, 1982. A public footpath follows the river along the left bank from the Barton all the way to the next road bridge where a minor road goes west to North Tawton.

Small bridges and some footpaths follow the waterway north to Bondleigh where a straight road leads via Taw Green to the river. Below Taw Bridge the Bullow Brook from near Winkleigh joins the Taw via the ford at Brushford Barton, meandering by some small but pleasant woods to the main Exeter to Barnstaple road and a sharp westwards curve along the valley, where the rivers Yeo, Dalch and the waters of Pepper Lake join it from Lapford and Zeal villages nearby.

Now we are at Eggesford where many acres of forestry plantations follow the road and river course, part of the vast Eggesford Forest complex of the Forestry Commission.

And on to Colleton . . .

Williamson writes of Tarka at Colleton or Colleton Mills, a pretty spot with its tiny hamlet where teas may be obtained.

The delightful moorland rough and tumble of the river has changed and the Taw sweeps in numerous meanders through much softer countryside. The river has now been joined by the Dart which divides Chawleigh and Chulmleigh and flows by Colleton as a place of herons and kingfishers.

In his "Survey of Devon" Tristram Risdon refers to "Coleton or Colstown House overlooking the Taw where the Dart drencheth itself in that river" and explains that it was the dwelling and ancient inheritance of the Cole family for many generations. Today the hamlet is visited more for a small firm producing handmade wooden furniture of excellent quality, than for any other reason. The little three-segmented bridge carried the old coach road to Barnstaple via Chulmleigh and Burrington and dates from the 17th century.

This is fine farming country. Trains, diesel now and not steam as in Williamson's days, still rumble over the girdered iron bridges, and jackdaws still nest beneath the criss-crossed metalwork.

At South Molton Road and the Kings Nympton railway station area we find Junction Pool with a heronry nearby where the great grey birds may be observed perched in the conifers. Kings Nympton, or Nymet Regis, because it was once part of the King's demesne, is some way up the road from the railway station. Indeed villages must have almost needed a train to reach the mainline train! The village itself is a conservation area with lovely thatched cottages protected by the planners, or should that be from the planners?

But Tarka stayed in the great outdoors and where better than Junction Pool, a delightful spot where the rivers Taw and Mole meet.

Kings Nympton Park Estate is now in the ownership of the Wildfowl Trust, better known by association with Sir Peter Scott and Slimbridge Reserve, but the estate is private farmland and not a wildfowl reserve. It is a carefully managed area of tenanted farms and cottages so keep strictly to public rights of way as signposted. This is excellent salmon fishing country where fishing rights have recently sold for thousands of pounds for a few

Junction Pool

Sandra Yeo

River Taw at Umberleigh   *Sandra Yeo*

hundred yards of riverbank. But if angling is not your game but a tasty trout is, then visit the trout farm nearby, where you can more less choose your own freshest of fresh Rainbows.

From the beauty of the dragonfly-sparkling Junction Pool, Tarka moves on to the third railway bridge over the river, wild country between Portsmouth Arms and Umberleigh with its many acres of forestry stretching along the river.

The old river bed Williamson refers to is close to Weirmarsh Farm and to the north of the railway. The road from Umberleigh to Warkleigh goes directly past the spot and the area is easily viewed from the roadside.

Otter Hounds often hunted this area, the field sport being quite popular hereabouts. The Inns at Portsmouth Arms and Umberleigh must have been very much a part of that scene in Tarka days and are so with tourists, anglers and locals today.

... "the last railway bridge above the tide" is downstream of Chapelton, a small hamlet best known for its sawmills and the making of wooden farm-gates seen throughout the countryside. This firm is owned by a Barnstaple timber company we meet with again on the river Yeo of that town.

Chapelton was once renowned for its fields of wild daffodils, locals coming here to pick them by the thousands to hawk them for sale in nearby towns. The greedy overpicking in the past has left the area almost denuded of this beautiful wildflower, sadly much the same story as elsewhere in the county.

The Taw at Chapelton pushes on swiftly as if eager to taste the salt of the estuary and feel the twice daily surging of the tidal ebb and flow. Lovely otter country this. Kingfishers and herons enjoy the fishing and even cormorants, so strongly associated with the estuary and sea coast, penetrate well inland to these fresh waters. They are commonly to be seen perched high in the trees even beyond Umberleigh, watching the great river as it sweeps seawards between the patchwork of fields so typical of North Devon's farming country.

We now reach the point where Hunt followers park on a road-bridge and White-Tip, the bitch otter arrives after a three hour chase by the hunt.

This is Newbridge close by Bishops Tawton village and a good place to pause to watch the river for signs of wildlife. More likely one will see mink here for it is a favourite haunt of these darkly chocolate-coloured animals, bold creatures often about by day. The bridge is more attractive from the field upstream than it appears from the road, with a lovely view downstream through its arches.

Bishops Tawton is a thriving village of fair size, yet unlike some growth villages it has retained much of its character. Bishops Tawton's name is due to its being the first seat in the county of Devon of a Bishop, the Bishop Werstan sent to Devon to sort out our "lack of ecclesiastical discipline" by Plemond the then Archbishop of Canterbury. It is not recorded whether it worked … even to this day! (Tawton, on the Taw). Footpaths follow the river and are ideal from which to view Tarka Country, the Taw here being a magnificent sight as it sweeps grandly on towards Barnstaple.

From Chestwood Hill on the Barnstaple side of the village fine views of the Taw can be obtained along the roadway. Across the river and lying some way from it is the village of Tawstock with its interesting church. Tawstock Court was once the home of Henry Lord Tracy, a Baron of Barnstaple and Judge in the reign of Henry the Third.

The church no longer has its ring of bells but, like Bishops Tawton, the village retains its character, being picturesque and a useful place from which to wander the left bank of the Taw via Tawstock Woods footpath.

Henry Williamson tells us of the Otter Hunt here and of the hounds. Dewdrop "the only true otter hound in the pack", of Render, Fencer, Hurricane, Barbrook, Bellman, Chorister, Waterwitch and others – and too of Bite'm the terrier – names that roll from the tongue as did their very baying when on trail of otters.

We are now on the upper reaches of the estuary with fine woods along the banks where we find Spady Gut not far below Newbridge. To the locals of the Barnstaple area, Spady Gut is well known as a little creek emptying into the Taw. To the visitor it may seem unimposing but it is as pretty a stream as any to be found and, of course, food and shelter habitat for aquatic or semi-aquatic mammals.

Tarka moves on from Spady Gut to Barnstaple Bridge, a sort of "gateway" to the estuary proper.

Records show that there was a bridge at Barnstaple in the 12th century but the present structure, widened in recent years, dates from the 16th century.

The bridge over the Taw river at Newbridge, nr. Bishops Tawton Graham Madge

Barnstaple Bridge Sandra Yeo

The bridge has fifteen arches and is a popular spot for anglers who fish the Taw for flounders, mullet and bass with lines cast from the bridge walling, as do the cormorants commonly seen diving below.

Williamson refers here to the railway viaduct curving across the Taw immediately below the road bridge, but this was demolished in the 1970s following the closure of the Barnstaple to Ilfracombe line.

Otters frequently use these intertidal waters, some moving from the Torridge to the Taw as well as frequenting the coast, particularly in winter when the sea offers good fishing. Their great preference is for eels and they must do quite well on this diet for such are common enough in the North Devon rivers and tributary streams.

As well as rod and line angling, much netting is done from boats and land. Pranging, the awful method of fishing with a pole with many metal spikes on a bar at one end rather like a rake, is also carried out. The user wades up and down at low water thrusting the pole into the river bed, a quite indiscriminate and cruel way of fishing. Night lines are also commonly used. The nightliners have metal rods poked into the sand at regular intervals each having a line stretched between them carrying dozens of carefully baited fish-hooks. The fishermen then meticulously sand over each baited hook to disguise the bait.

"Don't want the birds to know what's on 'em do I?" one said and of course the bait would otherwise have been a useful larder for opportunist gulls before the tide covered the lines. After the next ebbtide the fishermen revisit the lines to see how their hard, very cold work has reaped them a harvest. Men of the river indeed.

From a conservation point of view much of the estuary is now protected. From just below Barnstaple Longbridge and downriver to Bideford, taking in some of the tributary areas, the estuarine habitat is a Site of Special Scientific Interest (SSSI), a Nature Conservancy Council designation of protection. This is mainly because the area is nationally important for its birdlife. Williamson would have been pleased for he was concerned that we do not do enough to protect our environment. "Far too much talking and far too little doing" he said on more than one occasion.

Tarka entered the Yeo from the Taw on a rising tide, chasing mullet and investigating "broken kettles, cooking pots, basins and battered oil drums thrown away in the mud". Fifty years or more has seen little change in man's

Yeo river at Barnstaple

The Mill Leat

Sawyers bend on the Barnstaple Yeo

The "old lime-washed mill" near Derby, Barnstaple with a glimpse of the River Yeo on the left.

habits in that respect and rubbish is still dumped into our beautiful rivers. The mouth of the Yeo in Tarka's day was between fields and trees on one side, Monkey Island and Barnstaple Town on the other.

Now where 'Monkey Island' was is the ten-storey Civic Centre, the home of local government for North Devon.

And Tarka swam after mullet, passing "weed hung stakes leaning over the mud glidders". They remain, mossy and green as part of the riverbank retaining 'walls' around the perimeter of Pilton Park.

When Williamson wrote "Tarka" the main visual difference would have been huge piles of logs along the riverbank, great tree trunks waiting the saws of the timber mill. They were the dens of our younger days for the piles of logs had huge caverns within, ideal hiding places for children, and for otters. Now the logs have been replaced by rows of neatly cut planks and the saw-mill use of the riverbank close to the town helps reduce the visual harshness of concrete and steel.

The timber yard is owned by the same company operating Chapelton sawmills met with earlier on the Taw. Workmen there tell of numerous sightings of mink and just occasionally the sighting of an otter moving upstream of an evening.

More often it is their footprints one finds in the mud near the two huge drains emptying the mill leat into the river here near Rolle Bridge.

The "narrow gauge railway" referred to in "Tarka" would have been the old Barnstaple to Lynton railway which closed in 1935. It is possible to follow its route via the road and to see the massive viaduct at Chelfham still spanning the road to Stoke Rivers. Tarka chased mullet to Swine Park, out of the tidal water and into fresh and here we find the old limewashed mill standing between the Yeo and the leat built to turn its waterwheel. The remnants of the wheel are still on the mill wall as is the old "iron bedstead fence" Williamson mentions. Built to last, the old beds.

The Mill is well conserved and is a sausage skin factory, a link with Swine Park to this day. Upstream of the mill is a weir where trout may be seen leaping near shaded well-wooded riverbanks. Once really rural, certainly in Tarka's days, the area is now being gradually encroached upon by modern unattractive housing.

Close by is Derby, an area of homesteads on the fringe of Barnstaple. Once notorious for its toughliving people, they were and are today some of the warmest and kindest, the first to flock together if anyone is sick or in need of help. Today the Derby area is vastly different from the days when Williamson wandered there with hundreds of redbrick terraced houses demolished as "slum" clearance and replaced with modern Council estates.

I talked to some older characters of this area, men and women in their seventies and eighties, and some had seen otters regularly from the Pilton Causeway and along the Yeo hereabouts.

"Used to watch 'em come up river afore sunset, over the bridge we'd watch 'em" said one old man.

"They wud'n rare back in those days before the War and even up to the 1950s we'd watch 'em. Me'n the missis used to sit opposite the timber yard – huge logs there then not the scraggy planks you get there now. The otters would play about on a summer's night and chase the trout under the arches." His wife nodded. "Oh yes, my father watched 'em up by Derby weir chasin' eels too. Nought special to see an otter back then – old Williamson 'ad it easy in his day."

Probably Williamson did for though otters were in all likelihood never very common, they are well remembered by many country folk in days when there were certainly no mink about and there was no argument about identification.

Tarka moved swiftly away from Barnstaple to Swimbridge never knowing the interesting "otter link" between Pilton and the fair-sized village. Just beyond the area of Pilton known as The Rock is Higher Raleigh Road leading on to the North Road out of Barnstaple. Here is River View, a row of houses which were formerly the Otter Hound Kennels of the Cheriton Hunt and once the home of Deadlock and co.

The kennels were converted to dwellinghouses some fifty years ago not long after Williamson had written "Tarka". Indeed one of the present occupants of River View tells me that sixty years ago he would stand and watch the hounds being taken to hunt on Saturday mornings. "Great shaggy dogs some of them and some were foxhounds" he told me. "I was always amazed how the kennelman knew them all by name – they all looked the same to me", he said, "but mind you I was only ten at the time."

Henry Williamson would have known well the links with the two areas and I could not help pondering

Pilton bridge over the River Yeo

whether there was any significance in Tarka's movements directly from Pilton to Swimbridge. An unanswerable question.

Tarka arrived at Swimbridge via a brook running into the Taw, arriving under Mazzard orchards to find an old quarry. The quarry is still there of course but the mazzards, or wild cherry, are a rarity. Almost as endangered a species as the otter, the mazzard is only now being "revived" by local nurseries as a fruit tree with a future, a most attractive tree for any small garden, producing fine blossoms and fruit for ourselves or to attract birdlife.

Swimbridge is well known for Jack Russell; Parson Jack Russell the man, the Jack Russell breed of terrier and, too, the Inn bearing this name in the heart of the village.

This very tough little breed of terrier is extremely popular as a real country dog and may well have been the breed used in conjunction with hounds during otter hunts, sent into the otterholts to force the otter into the open.

Tarka hunted the deep pool in the limestone quarry, finding no fish and moving on found another drowned quarry where carp dwelt. He chased these to no avail and, hungry, he followed the brook away from Swimbridge to Exmoor.

Exmoor in summertime, "when the bees' feet shake the bells of the heather" says Henry Williamson and he briefly philosophises about hunting for sport. He talks of man's attitudes to other creatures and of pity acting through imagination. "A rainbow may be beautiful and heavenly, but it will not grow corn for bread" he says.

True enough but rainbows are a natural product of sun and rain as is the corn, the one feeding our need for beauty to compensate for the ugliness in the world, the other our need for food. Williamson, I know, hungered for both in his day. But Tarka arrives at Pinkworthy Pond amongst Harts Tongue ferns and the sound of croaking frogs . . .

Pinkery Pond *Sandra Yeo*

When the bees' feet shake the bells of the heather, and the ruddy strings of the sap-stealing dodder are twined about the green spikes of the furze, it is summertime on the commons. Exmoor is the high country of the winds, which are to the falcons and the hawks: clothed by whortleberry bushes and lichens and ferns and mossed trees in the goyals, which are to the foxes, the badgers, and the red deer: served by rain-clouds and drained by rock-littered streams, which are to the otters.

The moor knew the sun before it was bright, when it rolled red and ragged through the vapours of creation, not blindingly rayed like one of its own dandelions. The soil of the moor is of its own dead, and scanty; the rains return to the lower ground, to the pasture and cornfields of the valleys, which are under the wind, and the haunts of men.

PINKWORTHY POND (Pinkery)

Exmoor and the tarn lying under two hills and a bog called the Chains, a fabulously eerie place and a must to visit for all Tarka fans.

Williamson's story invariably comes to life at Pinkery. Deer slots in the mud, a herd of Exmoor ponies grazing the northern end and ravens cronking overhead are all common sights, whilst on some visits the pond heaved with hundreds of frogs during the breeding season. One of the largest otters I have seen was here at Pinkery on a rainy day in July 1980. The animal moved silently through the water leaving a wide 'V' wake behind it and then at our approach dived and reappeared several times until it eventually became hidden in the reed fringe and was lost to sight. We found its tracks around the pond edge and in one place at the south end of the pond, two otter spraints. An exciting day.

Williamson's reference to the pond as a tarn has always interested me for tarns, surely, are natural; ponds man-made. Pinkery is man-made and was probably dug to provide water for farming. Some say it was made for mining operations in the area, others as a lake to be enjoyed. Certainly it is the latter now. Of course there is a strong likelihood that there was a natural tarn there originally and the man-made merely an enlargement of it.

The Chains is a wet-underfoot place akin to the Great Kneeset on Dartmoor. Desolate and even treacherous when mists fall swiftly, it is no place to be caught out in

Exmoor rain which can be six times as dense as elsewhere, hence the saying "Exmoor the teapot, Tiverton the spout".

The pond is supposedly haunted by the ghost of a young farmer who drowned himself there in 1880. The age-old belief that a candle floated on a loaf of bread comes to rest over a dead body failed to find him, as did divers brought in specially from Wales. Eventually the pond was drained and the body discovered close to the bank in shallow water.

HOAR OAK WATER

Hoak Oak Water is a cheerful waterway, and along it one finds the famous Hoar Oak tree, a tree replanted several times over the centuries.

The first Hoar Oak we know of fell dead of old age around 1658. History tells us that about four years later another was planted in its place. This may well have lived until 1916, for then it was recorded that a Hoar Oak fell and was again replaced. This surely is the sapling "growing near the old stump of its father" as Williamson refers to it.

The tree is an important landmark and boundary tree between Brendon Common and that of Lynton and, too, the Exmoor Forest.

Another famous oak, the Kite Oak, once stood high on the lovely Chalk Water near the spot now known as Kittuck.

The Hoar Oak Water itself begins in a dripping goyle in the Chains Valley and follows a somewhat lonely course over the moor to the Lyncombe Woods and then a further three miles to Hillsford Bridge and Watersmeet above Lynton.

This is dipper and wagtail country with red deer and an abundance of wildflowers including the sphagnum bog-loving Sundew. Here a rough track crosses Hoar Oak Water near a ruined sheepfold with fine walks to Furzehill Common and Barbrook.

For the keener Exmoor walker the way to Hoar Oak Water is to follow the ancient boundary wall from Brendon Two Gates to Farley Water, then over Cheriton Ridge. The Hoar Oak Tree is close to where the wall crosses the water here. Hoar is almost certainly from 'Ore' meaning a boundary.

Hoar Oak Water   *Sandra Yeo*

As the water enters Devon from Somerset it widens a little to slow where it flows by a stand of mature beech trees. From here to Scoresdown Bridge the left bank is well wooded, then suddenly the whole waterway becomes beautifully shaded by the many trees, a sun-dappled place of great beauty.

Oaks predominate on both banks with hazel, holly and alder interspersed along the way. Here the water sparkles and tumbles its way to Hillsford Bridge where the Farley Water joins it.

For a further three parts of a mile Hoar Oak Water speeds to Watersmeet, to merge with the East Lyn river near the National Trust tearooms and shop.

A delightful waterway pouring into one of Exmoor's loveliest rivers, to this day a fine place for otters and their like.

Tarka finds White-Tip's scent trail here "where two waters meet to seek the sea together". He finds a drowned otter caught in a gin trap under the wash of the fall. In fear he moves on via Beggars Roost Hill to the lovely little

Ilkerton Water.

This Water runs off Ilkerton near Woolhanger and northwards to the Lynton road, pouring into the West Lyn river amongst lush ferns and mosses.

On a sun-dappled day in summer I came to explore this part of Tarka Country when Grey Wagtails were feeding their recently fledged young, their own yellow colouring seeming as moving flashes of sunlight upon the watersmooth stones.

Exmoor rain will turn these playful summer waters into sudden raging torrents and "the whirlpools that were Tarka's playthings".

Tarka followed the water to Lynmouth, hunted again by the Cheriton and fought with Deadlock, escaping via the sea at Lymouth.

Williamson is at his most exciting here and one breathes again as the otter releases his hold on the hound and "vanishes in a wave". The otter leaves this part of the coast and makes his way via East Cleave to the beautiful Heddon river valley, reputedly the warmest valley in England.

Tarka spent a week at Heddon living in the disused limekiln overlooking the Heddon Water. The kiln remains, a monument to man's lime-burning days as well as to a fascinating part of the "Tarka" story.

The last mile of the river Heddon's journey to the sea is between extremely steep valley sides with slopes so abrupt the vegetation is sparse indeed. With valley walls reaching to 700 feet the gradient is about 7 in 10 in places and the phenomenon of soil creep is much in evidence.

During his stay here Tarka witnessed a stag and three staghounds crash down the Heddon scree to be smashed upon the rocks. Ravens and Buzzards came immediately to feast upon the fresh carrion until the huntsmen arrived. Tarka slips quietly into the sea away from man and his dogs, that night "squatting on a rock" to eat conger. The west wind brings him the scent of White-Tip and the otter moves on to the area of the Great Hangman.

The sweep of hogsback cliffs from Heddon to the Hangman is quite magnificent with heights above the sea from 1000 to 1100 feet. These rocks are the famous Hangman Grits and consist of grey or red spotted gritty sandstone with some shale beds present. Beautiful countryside rich in wildlife with ravens and buzzards soaring amongst diving jackdaws.

When Williamson wrote "Tarka", buzzards were abundant, nesting in every wood as well as on the cliffs. During the 1950s and 60s we witnessed the drastic decline of raptors generally due to pesticide poisoning. Thankfully in recent years, due to the banning of the use of some of these toxic chemicals the birds of prey have made an amazing recovery and the buzzard is once again a common sight in the two river countryside.

In the quiet coves we still find Williamson's "shore rats" searching the tidelines for food. They are as different from town rats as are town and country house-mice, standing to eye intruders quizzically before scampering away to shelter in some hidden recess or cave in the cliffside.

Wild Pear Beach is today more strewn with human litter than weed as in Tarka's days, but it is nevertheless a lovely spot close to the sprawling garden village of Combe Martin. All can be viewed from the coastal footpath, the scenery from Great Hangman Look-out being an exciting panorama of cliffs and sea. One could see Tarka clearly in the mind's eye lolloping his way out of the waves and amongst the rocks and pebbles explored by Purple Sandpipers on passage through North Devon.

The sound of the woodlark's "wistful falling song over the bracken" is a rarity anywhere in Tarka Country today and probably was not at all common in Williamson's day. Decreasing severely in numbers in the past twenty-five years and hit by hard winters the woodlark is sadly a rare member of the North Devon avifauna. Williamson's reference to it in "autumn's little summer" is interesting in that woodlarks are known to flock at this time, small numbers and singles being observed along our coast and on Lundy during September and October. My own observations include eight woodlarks together on Great Hangman in January 1983 and some few and far between breeding records elsewhere.

It is also interesting that Williamson refers to "Phalacrocorax carbo" standing on the Morte Stone near Mortehoe. This "Isle of Wight Parson" is the Common Cormorant which frequents the coast as a breeding species and is seen well inland on lakes and waterways. Once Williamson pointed out one of these birds fishing at Ramshorn Pond and again used the Latin name. He said, when I stared at him in puzzlement, it was because "some bird person picked me up when I called it a Shag".

Shag is the very much used local name for cormorants

Sandra Yeo

Skirr Cottage, Georgeham

and the fact that we have the true Shag (Phalacrocorax aristotelis) also breeding on our coast sometimes causes confusion. The shag is widely known as the Green Cormorant anyway but is not a bird of inland waterways as is P. Carbo. Thus shag will continue to be used as a local name for both species for years to come I would think, just as crane is for the grey heron and dipchick for moorhen.

But on now to Pickwell Down where Tarka was disturbed by man, dog and ferret, all three of which still work the area for rabbits.

The area around Georgeham is attractive with its well-kept cottages and farms prettily clustered about the stream. The church of St. George merges well with thatched roofs and whitewashed walls, the locals taking a great pride in their appearance.

At Skirr we must pause awhile to see where Williamson lived for so many years. Tarka was at Skirr at Williamson's door and left his seals or footprints for him. A nice touch that the author brings Tarka here and to find that "Skirrrr" is one of the calls of a barn owl, here lifting off from the lopped bough of a churchyard elm.

The barn owl remains the emblem of Henry Williamson, the sign on the door of his home and now the symbol used by the Henry Williamson Society.

Cryde is Croyde, a large and popular village with fine beaches loved by tourists and locals alike. Cobwalled thatched cottages give the village an Olde Worlde charm enhanced by the old Devon farm buildings of cob at the top of the main street. Little has changed to spoil the place since Tarka's days though the elm trees, once a feature in every hedgerow, are sadly gone, decimated by Dutch Elm disease in the 1970s.

I found otter spraints at Croyde near the bridge over the stream right in the village — Tarka's descendants passing through the area when the inhabitants of the village lay asleep. Tarka leaves Croyde on an overland journey across farmland until he hears the roar of the surf and sees the lighthouse at Braunton Burrows.

Soon he is again at Ramshorn Pond and the area of Horsey Weir and Horsey Island.

At Braunton Burrows one night in summer when the dunes were deep purple shadowed and the sun a mere dull red strip of light upon the sea horizon I talked with a Water Bailiff who works the area. His very job takes him amongst the wildlife and he spoke of sightings of otters rising from the waters lapping the beach and lolloping inland towards the wet-slacks and marshes.

No longer though do the lighthouse beams shine upon the wet sands at Crow, one fewer sign of man upon these lonely estuary shores. A lighthouse was here until very recently but was demolished as being obsolete and too costly to maintain. Its last occupants of note were kestrels who regularly nested upon the structure in springtime.

Braunton Burrows is in part a Ministry of Defence Training area, the Nature Reserve subleased by the MOD to the Nature Conservancy Council. The MOD lease the Burrows from the Christie Estate Trustees, the Christie family having long been part of the North Devon scene. Visitors to this part of Tarka Country have the choice of two free carparks, one at Sandy Lane near the American Road and the other at Broadsands via the Toll Road.

The "American road" is so named because American troops trained here for the Normandy landings of World War Two. There are still too many relics of those days strewn about including a lessening number of unexploded shells. One such was regularly used as an anvil by a Song Thrush who smashed snails upon its surface with a gusto born of innocence of such terrible weapons.

(Never touch suspicious looking objects in such places but report them immediately to the police, marking the spot with a stick pushed into the sand). Here at last Tarka finds White-Tip again and with other otters they hunt salmon and other fish in the estuary, playing with cubs on Crow Island and with sticks and the empty shells of skates eggs.

They swim out to the Shrarshook as curlews wing their way landwards, eating mussels in the pools as salmon "ran" up the fairway on the tide flow "feverish to spawn in the fresh waters of their birth".

Williamson says Shrarshook or Sharshook depending on the amount of beer the fishermen and deep water sailors had drunk, a Captain Charles Hook being caught by the tide here one night and drowned. This estuarine gravel ridge is today better known as Middle Ridge. After a time of hunting eels and many mullet Tarka and White-Tip leave the cubs at Horsey Marsh to begin their own lives, the two older otters moving back to the Torridge, the river of their birth.

Near Canal Bridge they meet with a Loon or Great Northern Diver, its beak "sharp as a rock splinter".

It is in such ways throughout his story that Williamson tells us the time of year for in North Devon this bird is a winter visitor. Quite often it is seen in the estuary and inland on fresh water, early birds arriving in September and observed in the area into April and May. Thus as well as being a beautiful story "Tarka" is a fund of information for the country-lover and naturalist.

Tarka and White-Tip come back to Lancarse Pill and the area where they first met when she was travelling with Greymuzzle. Their travels have brought them full circle to the salmon-filled river though today salmon numbers are declining rapidly, due to a variety of factors including pollution and over-fishing at sea. Whimbrel are also mentioned here, this elegant curlew-like bird being seen on autumn and spring passage in North Devon. Its rapid seven-note call may often be heard on moonlit nights as scores of the birds fly over as passage migrants, either from the south in spring, the reverse in autumn.

Tarka and White-Tip pass a happy winter on the Torridge and with the spring came the River Martins, as Williamson calls the Sand Martins, returning from Africa to summer with us.

At Twin Ash Holt cubs were born to Tarka and White-Tip, Tarka visiting his mate every night from the Pool of the Six Herons. March moves into June, young herons coming to fish the Torridge with their parents. Old Nog with his family hunting bass and eels, and White-Tip brings four cubs onto the river.

Summer is spent joyfully fishing and playing in the area of the holt and Canal Bridge.

Here the otters enjoy the beautiful, deeply green pools above Beam Weir, playing beneath the bridge arches and the shady tree-lined banks of the Torridge, or below the weir in the fish-filled shallows and grassy damselfly-lit fringes. Tarka often fishes with his eldest son Tarquol here, and it is to Canal Bridge he comes to play "one last game" with White Tip and the cubs before moving on alone as is the way of adult male otters . . .

> *The sun looked over the hills, the moon was as a feather dropped by the owl flying home, and Tarka slept, while the water flowed, and he dreamed of a journey with Tarquol down to a strange sea, where they were never hungry, and never hunted.*

And we read of Town Mills between Bideford and Torrington as the scene of the gathering of various West Country Hunts for the annual "Joint Week" of early summer. A great social event, this, in the Hunt calendar, the many-hued uniforms "coloured as the dragonflies over the river" and watched by many locals.

Tally Ho!

The hunting of Tarka began shortly after ten-thirty in the morning, twenty-three hounds and two terriers with many men and women following, all eager for the chase and the finding of an otter to pursue to its death.

It is Deadlock the hound who finds Tarka's scent where six hours before the otter had touched the shillets. The Hunt followed his trail across meadows to the weir, the otter having cut the corner of the horseshoe meander of the rivercourse and left his scent on the grasses.

They follow Tarka's trail to a holt upstream of the weir, battering at the soil above it with a crowbar until eventually a "chain of bubbles" betrays the escaping otter fleeing into the mill leat.

Relentlessly followed by the Hunt, Tarka moves into Dark Hams Wood and is again pursued back to the leat and to the mill itself where he hides in the topmost trough of the mill wheel, stilled now during the mill-workers lunch break.

Tally Ho!

At two the men return to work and Tarka is thrown into the river as the millwheel once more resumes its turning. The otter flees to a pool below the bridge, then on again, the Torridge a melee of humans, dogs and the purposefully wielded poles of the stickle.

Hunted to Taddiport Tarka passes under Servis Wood, fleeing on to Elm Island where the terriers catch him and they fight until a fall into the river enables Tarka to escape yet again. He moves back to Beam Pool where he finds his son Tarquol. They rest together for a while, but again the Hunt is upon them and the Huntsman begins his steady pounding of the holt-top with the iron bar.

Scenes taken during the filming of "Tarka" on the Torridge near Weare Giffard, kindly lent by Mrs. D. Bond.

Tally Ho!

    The two otters move out into the river and it is Tarquol the hounds chase after, hunting him to Furzebeam Hill where he is caught by Deadlock. We read of the cheers of the "sportsmen" as Tarquol, blinded now and with his jaws smashed, is killed. "He has gone home before Tarka"...
    Tarka meanwhile, discovering that the Hunt has not followed him, swims on to Beam Weir and to the Kelt Pool where he is found and hunted yet again.
    In the sixth hour of hunting he eludes his pursuers, hiding in peace in quiet water until a dragonfly alights on his nose and his sneeze betrays his presence to watchers on the riverbank. Bitten and biting in return, Tarka is hunted to Peal Rock and in the ninth hour is so fatigued he drags himself from the river. Beaten by stickle poles he fights with his remaining strength until Deadlock catches him by the tail and they battle into the water.

Tally Ho!

Tarka, bleeding from many wounds, escapes once more and in the tenth hour of the hunt the tide is at flood and the hounds are called off. As they are about to leave, Tarka is spotted "moving with the tide, his mouth open" as he drifts exhausted with the current. Deadlock jumps down the bank, biting Tarka in the head and throwing him, falling with him into the river where the otter bites into the hounds neck as they sink into deep water. On a changing tide that began its ebb to the sea, Deadlock and Tarka fight beneath the surface of the river close to the otter's birthplace. Deadlock the hound dies in this last battle and here are seen the great bubbles of air that are the last we read of Williamson's immortal otter.

And while they stood there silently, a great bubble rose out of the depths, and broke, and as they watched, another bubble shook the surface, and broke; and there was a third bubble in the sea-going waters, and nothing more.

The Cheriton Hunt

Otter hunting on the Two Rivers Country was principally by the Cheriton Hunt. The Cheriton's territory lay entirely within the County of Devon, the Cheriton water consisting of the Rivers Taw, Torridge, Teign and all their tributaries as well as the Creedy and many other clear, swiftly flowing Devonshire trout streams.

The pack was started in 1846 by William Cheriton, who was Master for thirty seasons. From 1876 the Hunt knew several Masters and was the first Hunt to have a lady Master, a Mrs. Beaumont. The kennels were situated at Pilton.

Records show that some fifty otters per season were killed by the Cheriton in the years up to 1940 with an average of nearly twenty kills a season after that time.

The counties of Somerset, Devon and Cornwall can be said to be the home of Otter Hunting and history tells us that in the first years of the thirteenth century, King John sent twelve couples of his hounds to the Sheriff of Bristol. These arrived with orders to hunt otters in the Somerset streams for two months "at his own proper charges".

The Westcountry packs were the Culmstock, the Dartmoor, the Cheriton and the Tetcott.

Otter Hunts

In the last chapter of "Tarka", Henry Williamson refers to the presence of Hunts other than the Cheriton, out for the kill during "Joint Week". This took place in the early summer each year.

He refers to the Culmstock, the Crowhurst, the Dartmoor and the Courtenay Tracey and Eastern Counties.

The Dartmoor Hunt

Founded together with a Foxhound pack in 1824, and known first as The Lyneham, then as Mr. Trelawney's, the Hunt became the Dartmoor in 1874.

All its rivers flow into the English Channel and are smallish trout streams always known for their otter populations. For many years the Dartmoor pack consisted of mixed Otter and Foxhounds. Records show an increase in kills as this hunt progressed through the years.

The Culmstock Hunt

Begun by a Mr. Jewell Collier about 1790, he holding the Mastership until 1831. Mastership remained in the Collier family until 1898, with several subsequent Masters.

The Culmstock hunted the dykes and marshes of Somerset — Athelney and Sedgemoor and into Devon, with its principal river being the Exe.

Keen rivalry existed between the Culmstock and Hawkstone as to which could account for the greatest number of kills in a season, the Hawkstone holding the record with seventy otters.

True otterhounds were used, but the Culmstock gradually became entirely a foxhound pack and the purist "Otterhounds for Otter Hunting" brigade were taken over by the "Foxhounds for Everything" school.

The Courtenay Tracy Hunt worked the chalk stream countryside of Hampshire, Wiltshire and Dorset. Founded by a Mr. Tracy who was Master for twenty-seven years, the Hunt averaged over twenty kills per season, again with an increase of kills "after the War", (World War I).

After Mr. Tracy retired, his long friend and henchman, Captain Arthur Hussy, took on the Mastership for a further twenty years until 1934. Hounds were mixed, with foxhounds predominating.

The Crowhurst Hunt

Formed to hunt the counties of Kent, Sussex and parts of Surrey and Hampshire, the Crowhurst was founded in 1903, introducing the sport for the first time into this part of England.

Much of the hunting was ditch-hunting, there being few rivers of note in the Crowhurst territory and most streams being deep and steep-sided.

The pack was mainly of smooth hounds with a few Otterhounds and several cross-bred hounds. Unlike most other areas, the Crowhurst territory did not yield more otters following the War and the average fell to ten a season suggesting an early and dramatic decline in those counties.

The Eastern Counties Otter Hounds

The Essex rivers were hunted by a pack purchased from the Culmstock in 1897 and eventually the E.C.O.H. took over all the rivers of Norfolk, Suffolk and Essex plus some in Cambridgeshire and the borders of Hertfordshire.

The country, being a mixture of fens, carrs and wide expanses of water, plus extensive saltmarshes, was good otter habitat. Kills of thirty per season were recorded in the 1930s and this Hunt held the records for killing the largest otters known, dog otters of 35 1lbs and 34 1/2lbs.

The pack consisted mainly of rough hounds with a few smooth hounds present.   Note: By arrangement with the E.C.O.H., The Bure Valley Otter Hounds, founded in 1927, hunted a part of Norfolk including the Bure, Yare and Wensum rivers.

The main Tarka characters

Just how good a naturalist was Henry Williamson? Is it even important, for he was a brilliant enough writer?

I feel it is for there are those who say he wasn't and disagree with some of the incidents within the story.

Without question Williamson was an excellent observer. He followed every inch of Tarka country thoroughly and each boulder and tree was as he wrote of them — many still are.

Let's look first at some of his characters, at Old Nog the Heron, Kronk the Raven, Swagdagger the Stoat, Fang-over-lip the Fox, and others, and examine, too, a few of the incidents they were involved in.

The Great Winter

While in the Braunton area Tarka and Greymuzzle experienced a winter that could only have been equalled in later years by those of 1947 and 1962/3. My own memories of both, 1947 somewhat faded, but 1962/63 being most vivid, are such that I well recall the awful toll they took of wildlife.

One incident in particular has been the cause of much argument, the killing of a Mute Swan by the two otters. Personally I am sure this could easily occur, certainly more easily in water, for here otters are in their element and the swan, though a powerful bird, is but a paddler upon the water and out of its element beneath it.

> "... and then came a north wind which poured like liquid glass from Exmoor and made all things distinct. The wind made whips of the dwarf willows, and hissed through clumps of the great sea-rushes. The spines of the marram grasses scratched wildly at the rushing air, which passed over the hollows where larks and linnets crouched with puffed feathers. Like a spirit freed by the sun's ruin and levelling all things before a new creation, the wind drove grains of sand against the legs and ruffled feathers of the little birds, as though it would breathe annihilation upon them, strip their frail boines of skin and flesh, and grind them until they became again that which was before the earth's old travail. Vainly the sharp and hard points of the marram grasses drew their circles on the sand: the Icicle Spirit was coming, and no terrestrial power could exorcize it.

We must remember that, starving though the otters were, so must the swans have been also.

In 1963 I picked up mute swans from the banks of the Taw, their bodies emaciated to an unbelievable degree, the skin and feathers hanging loose from their bones. The allegedly fearful wings beat but feebly, only the nip from their bills being painful.

At Barnstaple Longbridge the Taw froze solid, and we walked across, so bitter was the cold.

Knowing mink can tackle and kill a Canada Goose defending its nest and young, and also kill peafowl I am absolutely certain an otter could take a swan under these conditions.

The "arm-breaking" stories of a mute swans' wing beats refer to possibilities if a small child was attacked. I have caught and held several swans that have had fishermen's traces and hooks attached to them and have yet to find one too difficult to capture and hold, apart from the obvious problem of approach if the bird is on the water.

Thus I firmly believe that this incident could occur given the circumstances so carefully set out by Williamson. Maybe he actually saw it happen.

As to the fox, Fang-over-lip, disputing the kill and the badger, Bloody Bill Brock, eventually winning it, such are the ways of the wild, particularly when cold and hunger is rife. I've seen fox and badger reach flash point disputing water during a drought; the noise and the obvious rage frightening to the onlooker.

In winter when food is scarce even the meeker creatures, by our terms, will take on and win against the normally more agressive species.

But on to other things. One day, just for fun, I hid in reeds and as gutturally as possible croaked frog noises out across the water – as the raven did at Pinkery. To my delight a number of frogs came to investigate, and somehow I could well imagine Williamson testing this theory. And ravens? They are very intelligent birds.

The intelligence of wild animals should never be underestimated. Williamson tells of the badger rolling on gin traps to spring them. The poacher with whom I wandered the countryside as a boy told me similar stories and he, like Williamson, knew his stuff from first-hand experience.

In the late 1970s I spent many hours fox-watching, and learned a great deal about the intelligence of this animal. A fox at Braunton Marshes exploited a wildfowl shoot by taking mallard that fell, shot, into the reedbeds. This certainly happened on three occasions, no doubt the first happening as a "useful accident" and the fox learning to its advantage.

So read Williamson's nature stories carefully — as a naturalist if you will but they'll take some arguing with — much better to enjoy them.

The Otter

Now afforded the protection it deserved several years ago the Otter is no longer hunted with hounds, the sport having been banned.

Anyone suggesting that otter hunting had no effect on the animal's numbers and that the sport did not contribute to the otters' decline must be speaking without thought.

The continuous killing of animals which are never in very high numbers cannot be helpful to maintaining a viable breeding population. Hunting, together with the more modern problems of habitat loss and pollution, have together brought the otter close to extinction, the last two having hastened the decline.

North Devon has a breeding population of numbers that lends some hope for the otters future though the situation is extremely serious and the otter is still very much an endangered species. That they are hanging on in their stronghold areas, including North Devon, is because of our relatively unpolluted waterways, the retention of woodlands and other bankside cover along river and streambanks, and the fact that much otter habitat is privately owned countryside with few publicly disturbed areas.

Otters are quite large animals and usually measure three to four feet long, some dog otters having been recorded at five feet in length.

A typical member of the Mustela or weasel family, its bounding, lolloping gait on land resembles that of the weasel, stoat and badger. The colour is a uniform brown except for the white throat.

The five-toed feet are webbed, the legs short, the tail thick at the base and tapering away to a point.

The otter's head is flattish and somewhat cat-like, with the main sensory organs, the ears, eyes and nose set along the top of the head in a way that allows the animal to see, hear and scent whilst swimming with the remainder of its body submerged. Its hearing out of water is excellent but ears and nostrils are closed when under water. Eyesight is very good beneath water.

Otters live almost entirely on fish, crustaceans and aquatic insects. Some small mammals, and occasionally birds, may be taken during hard conditions.

Young may be born at any time of year but most cubs I have seen are about in summer and autumn, suggesting spring births are the more usual. Cubs born in the spring may remain with the mother well into early winter before dispersing to find territories of their own. Territories, however, are large and some otters travel miles between rivers, streams, estuaries and the sea coast. They will travel many miles overland between watersheds.

Most books state a territory of from five to seven miles but such a statement must be viewed as a possible average and no more. Otters vary a great deal just as do humans and their movements depend on suitability of habitat, food availability, weather conditions, disturbance, even whim . . . the latter being an important factor often disregarded by naturalists, proprobably because it is an immeasurable factor.

Ageing otters may remain in a secluded area for years until their death. I know of two such at the moment who are almost certainly in their very small, peaceful territories as I write. One of these has been in the vicinity of an old mill and leat for some eighteen months.

The otter's home is called a holt and is usually a cavern amongst tree roots, often reached by an underwater entrance. Here the young are born and raised. Occasionally holts are away from water, some high amongst trees above the waterways. Coastal holts may be in rock caverns or caves and in peat on moorland habitats. As with foxes and other mammals the otter may well have two ways in and out of its lair.

Outside the breeding "season" otters tend not to use the holts, having no definite permanent home and lying up by day in secluded sites such as reed beds or bankside shrubberies.

Stoats with rabbit kill — Belstone Cleave Dartmoor.

The Stoat

We meet Swagdagger the Stoat high on Dartmoor when he and Tarka argue over the rabbit kill the stoat has made.

The two animals are close relatives, members of the genus Mustela and Lutra respectively, the Otter being named Mustela Lutra by Linnaeus.

The stoat is fairly common in wild countryside and farmland, but like many wild creatures its numbers are declining. One Devon man I know calls the stoat a Hob, though I do not know where this term comes from.

The upper parts are reddish-brown, the underparts white, the tail black-tipped. In length from tip of nose to tip of tail the stoat measures 15-20 inches. Males are about fifty per cent larger than females.

Though it is a fact that stoats may turn white in winter (Ermine) this is not the case here in the south and the coat in winter is possibly a paler brown but nothing more.

As to habitat, I've seen stoats on the moors, in marshes, woodlands and farmland with the latter two habitat types being the more popular. This would be because they move carefully in cover and, of course, their distribution is related closely to this and the food supply.

The usual gait is bounding but when in scent-pursuit of prey the stoat walks with head low, absolutely intent on the pursuit. I've had them pass within a yard or two of me and have not been observed.

Stoats are good swimmers and climbers and I've seen them high in trees examining my nest boxes in North Devon woodlands. They are active at all hours of the day and night and can be attracted close if one imitates a rabbit's squealing sound.

During the breeding season, stoats keep company and males help to feed the young ones, but otherwise sexes live separately. Stoat dens may be in hollow trees, rock crevices and small caverns or burrows.

Rabbits and rats were stoats' favoured food, but with small rodents and birds more commonly taken following myxomatosis in rabbits. Fruit, insects and worms may augment the diet and some carrion is taken. I have seen birds mob a stoat in a country lane until the animal suddenly pounces on one, and believe they have actually learnt to exploit this situation as it is not uncommon.

ten young are born the following spring. Delightful animals are these.

The Fox

Fang-over-lip – what a fine name for the main fox character of Henry Williamson's story, conjuring up immediately a picture of a fox with a permanent grin on his face.

Foxes, fifty years after "Tarka", are still common enough, though today more than ever before they are persecuted by pelt hunters. Thousands are killed every year. A recent figure given by the Hudson Bay Company upon an enquiry from the Devon Trust for Nature Conservation was 100,000 pelts per year from Britain's foxes.

is ethically wrong and morally unjust. Farming and agriculture provides adequately for our needs and synthetic fur materials are easily manufactured and provide fuller employment.

Fox hunting is still much in evidence today as it was in Henry Williamson's "Tarka" days. More people follow the Hunt today however, the car being responsible in a way for this "availability" of field sports of this kind to more and more people. Hunting has become a more emotive issue in recent years and bloodsports feature quite importantly, even in some political arenas!

In the Two River countryside of North Devon, the fox is possibly the commonest mammal to be seen though they swiftly avoid contact with man, sightings usually being fleeting but always rewarding, for the fox is a beautiful animal.

That foxes live in earths is common knowledge, but earths may be well concealed or "open" above ground affairs, often in bramble brakes and the like.

Foxes roam far afield in search of food, rarely killing prey close to home unless times are particularly hard. I have observed foxes regularly leave their den areas to pass rabbits quite closely, the rabbits appearing quite unconcerned and continuing to feed as if they knew it was safer living close to the fox than away from it.

Largely nocturnal (except in more remote countryside where they will readily hunt by day) foxes spend much of the day in their earths. Earths are usually caverns in the ground or in hedgebanks but many North Devon earths are well made "burrows" with two entrances being usual. Most earths in my experience are made by the foxes themselves.

The fox's chief food items are rabbits, now that they are more plentiful again, rats, mice and voles, hedgehogs, squirrels, some frogs, snails and beetles are taken, plus considerable amounts of vegetable matter including fruit and fungi. Birds, particularly pheasants and wild duck, are also taken, as are sheep and poultry this being often a "local" occurence, a particular fox sometimes exploiting easier pickings and teaching its young to do likewise in some cases.

Foxes are fine parents and will courageously defend their cubs, males being very attentive fathers, playing with and teaching the, usually, four cubs of the single litter per year.

The Badger

Bloody Bill Brock is met with in the Braunton area of "Tarka" and indeed can be met with to-day in most of the well-wooded valleys of North Devon.

More completely nocturnal than most British mammals, the Badger lives a communal life in a sett, an underground home with passages, galleries, nursery etc. reached by a number of entrance holes. Cubs are born in the spring and usually number two or three.

Badgers in Britain do not hibernate, but are less active in winter.

Clean animals, they dig latrines away from the sett as well as changing their bedding regularly, both sexes assisting in these activities.

Badgers eat fruit, beech mast, acorns, grass, clover, roots, beetles, worms and slugs. Young rabbits are taken and voles and mice sometimes feature in their diet as well as carrion such as lamb carcass. They do not kill lambs. Occasionally a badger may learn to take poultry and may turn rogue if old age and infirmity forces it to turn to whatever prey is available.

In very recent years the Ministry of Agriculture has carried out a Badger Gassing programme believing the animal to be a contributary cause of Bovine TB in cattle. This has decimated the population in some areas and rekindled a feeling by some against badgers. Reports of an increase in badger baiting suggest the anti-badger brigade is on the increase and this terrible, disgusting practice along with it.

OLD NOG

The Heron

Old Nog the Grey Heron is the first named character to make his appearance in "Tarka" and is the very first creature mentioned at all. To-day Grey Herons frequent those self-same places of Tarka's birth and will be found in and along all the North Devon waterways as well as the lakes, reservoirs and ponds throughout the area. Wherever fish can be seen by their keen eyes there, too, will they be. Here the heron nests in trees, huge stick nests, some six feet across and often built by February, or old ones renovated to be used again for, like the raven, they are early nesters.

Tall birds of three feet or so in height, they are a delight to watch when fishing as well as in flight. Their slow, deliberate flight belies the power of the bird and they are able to cover long distances in a very short space of time.

Herons have flown parallel to our car at speeds of 25-30 m.p.h.

Quite common still they fare well except in very hard winters when many perish. They are colonial nesters and their heronries remain in an area for many years unless they are persecuted.

I have seen one or two changes in heronries in recent years and one heronry I've known well for some time was deserted because of too much human interference by school boys. Tree felling, of course, also causes problems and birds may move several miles when disturbance or loss of habitat occurs. Fortunately for the species, herons will nest in conifers or broadleaved trees.

Of many happy memories of herons, there was the late winter afternoon at Arlington Lake (National Trust) when more than thirty herons put up from the ancient heronry there to fly above the lake surface. This when viewed from the hide at close quarters was nothing short of awe inspiring.

The Raven

This magnificent bird, the largest of the Crow family, is a common sight in North Devon, and probably in as good numbers as in Tarka's days.

Kronk the Raven is one of the main characters of Williamson's book and he knew the bird well.

Ravens nest on cliff ledges, in quarries and in trees in North Devon, nesting early at traditional sites and with eggs laid usually about the first week in March.

On a day when misty rains swept in from the sea and obliterated Braunton Burrows from view, I wandered lonely cliff-tops to find Kronk the Raven's home.

Pink thrift flowers, the first of masses to come, brightened the rather cheerless day. The first wheatears scampered along the short, sheep-grazed turf, their feathers blowing up their backs in the chill wind.

Then the call of an adult raven, a deep-throated "pruck-pruck", and the great bird was before me on the cliff-top. Obviously agitated, it hopped along the cliff edge as I strolled, keeping close and turning its head on one side, its bright eyes watching my every move.

Was it leading me away from its nest in the manner of Plovers, I wondered? But no, suddenly there on the rainswept sandstone cliffside was Kronk the Raven's nest-site, and on the nest, three jet black young ones.

An exciting moment, for fifty or more years before this, Williamson had written of this very spot. Now the descendants of Kronk live on here.

Each year they rebuild and re-line the huge stick nest and raise their young in this precarious site above wave-crashed boulder-strewn sands, a superb place favoured, too, by rock pipits, stonechats and wagtails, who love the rocks and blazing gorse. Ravens mate for life and hold fast to these traditional sites. Legend has it that King Arthur did not die but lives on as a raven, one day to return in human form.

I would hope that the Raven in question is perched in a safer place than this when the transformation takes place!

The Barn Owl

A white owl features early in "Tarka" and throughout the story. Much commoner in Williamson's "Tarka" days when old barns were well used and not neglected as they are to-day, the barn owl had a wider choice of nest site and was thus successful.

As with many creatures, the 1950s and 1960s, took their toll of the owls and mankind has much to answer for poisoning the countryside.

The spread of modern farm buildings during and following this period did little for the barn owl. Even those searching for natural nest sites in hollow trees and such would be confronted with an unprecedented timber

THE WHITE OWL.
(BARN OWL).

felling programme which has drastically reduced the numbers of old mature trees.

Road casualties are also numerous and this great ally to the farmer has been treated badly by us in the past few decades.

Once a wild animal or plants' decline renders it an endangered species, or nearly so, its plight becomes recognised and mankind jumps into action. In the case of the barn owl, tea chest nestboxes are rapidly being erected on many farms to allow the birds the opportunity to make a comeback. These appear to be working very well - yet another sign that things can be put right.

Barn owls are often about by day though really they are nocturnal birds, hunting usually after sunset and into the night. During the breeding season when young ones need feeding and, too, in winter when food is scarce, the barn owls hunt a good deal by day.

Sometimes producing two broods in favourable years, the barn owls' breeding season can be long and hard particularly in bad summers. Incubation is about thirty three days by the female, and the period from hatching to fledging can be from nine to twelve weeks, some four months for one brood! The young are fed and cared for by both parents.

KINGFISHER.

The Kingfisher

Halcyon the Kingfisher enters Tarka's life when his eyes are about two weeks open.

I love the simple description of a kingfisher's flight where Williamson says the kingfisher "drew a blue line" – exactly as the eye sees this superb little bird in flight.

Kingfishers are fairly common along our slower moving waterways but they will never be numerous, too prone to hard winters. Their numbers pick up and are again decimated by snow and ice and, in their aquatic habitats, starvation.

Pollution, too, plays its part in the depletion of their numbers and the Taw and Torridge rivers, though cleaner than many, have their share of poisons. Never still for more than a moment, kingfishers are a joy to watch as they speed along the waterways with a shrill "peet-peet" call. Nesting in holes in banks close to water they lay six or seven glossy white eggs with both sexes incubating for some nineteen to twenty-one days. The young fly after three to four weeks of hatching and the sight of a family of kingfishers about the river is a sight to cherish.

In winter kingfishers move to the estuary and coastal habitats and may often be observed fishing in the tidal creeks of saltmarshes.

Conservation note. Like the barn owl, the kingfisher must not be photographed at its nest site without special Licence from the Nature Conservancy Council. There are several species so protected by the "Wildlife and Countryside Act" and a full list is obtainable from the N.C.C., 19 Belgrave Square, London.

The Buzzard

Henry Williamson refers to buzzards several times during "Tarka the Otter" and to one such, Mewliboy, in particular. Obviously referring to the hawks' mewing cry here the author would have observed them wherever he travelled in North Devon.

The area is regarded as a stronghold for the species for in Britain as a whole there are many counties where they are either non-existent or breed but sparsely.

Commonly seen as a large brown and whitish hawk soaring on widespread wings over woods and farmland, the buzzard is a sedentary bird and usually breeds each year in the same area.

A bulky nest of sticks is built by both adults usually in trees, sometimes on cliff-edges. Two or three eggs are laid in late April or early May and incubation takes about seven to eight weeks. Both parents tend the young until they fly at about seven weeks old when they often remain with the parents well into the autumn.

Buzzards eat small mammals, rabbits and rats, carrion and usually supplement their diet with earthworms, beetles and occasionally a few berries.

Sandra Yeo

Badger Books

The Heritage series looks at Devon as it is today and records our County — its buildings and its people, its wildlife and its rivers, its railways and its roads, its coasts and harbours, villages and market towns.

"Wildlife – Mammals" by Trevor Beer

A clear, concise description accompanied by the author's illustrations of the principal mammals to be found in the Devon Countryside, from Red Deer to Pygmy Shrews.

"The Cruel Coast of North Devon" by Michael Nix

Not merely a catalogue of shipwrecks, Michael Nix describes the agony of disaster and the efforts made to avert it from early lighthouses and lifesaving equipment down to the present day, ending with the wreck *Johanna* off Hartland Point.

"Back Along the Lines – North Devon's Railways"
– by Victor Thompson

Not written for railways enthusiasts alone, but a charmingly worded reconstruction of the history of the branch lines that did so much to improve communications throughout North Devon. Sadly, all save the line to Exeter have disappeared but fond memories linger on.

"VILLAGES OF NORTH DEVON"
"MARKET TOWNS OF NORTH DEVON" by Rosemary Anne Lauder

Of interest to visitor and local inhabitant alike, the author has travelled throughout North Devon and describes in "Villages" those with something to offer in the way of history, or picturesque appearance, or true Devonian character. In "Market Towns" Barnstaple, Bideford, Hatherleigh, Holsworthy, Okehampton and Torrington are in turn studied with a history and a modern profile of each town.

All the Heritage series' books are profusely illustrated with many charming old photographs and prints, and maps of the areas covered.

Priced at £1.95